The
LITTLE BOOK OF
PALAEONTOLOGY

Rasha Barrage

summersdale

THE LITTLE BOOK OF PALAEONTOLOGY

An Hachette UK Company
www.hachette.co.uk

Summersdale Publishers Ltd
Part of Octopus Publishing Group Limited
Carmelite House
50 Victoria Embankment
LONDON
EC4Y 0DZ
UK

www.summersdale.com

Printed and bound in Poland

ISBN: 978-1-83799-013-9

Substantial discounts on bulk quantities of Summersdale books are available to corporations, professional associations and other organizations. For details contact general enquiries: telephone: +44 (0) 1243 771107 or email: enquiries@summersdale.com.

Disclaimer
Neither the author nor the publisher can be held responsible for any loss or claim arising out of the use, or misuse, of the suggestions made herein. The publisher urges care and caution in the pursuit of any of the activities represented in this book. This book is intended for use by adults only. Please drink responsibly.

Contents

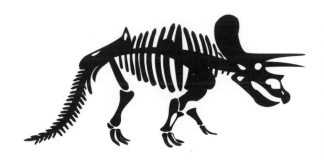

INTRODUCTION

Mental time travel is an incredible ability – we use our imagination to recall past experiences, picture the future and even transport ourselves to a different era without ever leaving our current location. Physical time travel remains in the realm of science fiction for now. But what if you could touch the past? What if you could see a creature that lived millions of years ago or discover a species that has lain hidden for thousands of years? For palaeontologists, this type of time travel is possible, and their discoveries are helping us to unravel the mysteries of Earth's 4.6-billion-year history.

Palaeontology is the scientific study of ancient life, including the evolution and behaviour of extinct organisms. Unlike archaeologists who dig up artefacts and architectural ruins to understand the earliest human cultures, palaeontologists focus on the fossils of ancient life – bacteria, fungi, plants and animals, including your childhood favourite: dinosaurs. This book takes you on a whirlwind tour of the subject, revealing our planet's distant past and the creatures that once walked where you do today.

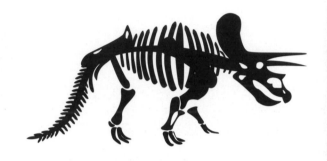

FINDING
FOSSILS

Over 99 per cent of all species that ever lived on Earth are now extinct. So how do we know about any of them? The answer lies beneath your feet – in fossils preserved in the ground. Humans have been finding fossils for millennia, but the scientific field of palaeontology was not established until the nineteenth century. This chapter will introduce you to some of the most important palaeontological techniques developed over the last two centuries, including the methods for finding, dating and identifying fossils. You will discover how these ancient relics allow us to reconstruct our planet's past and evolution, and why they are critical to our understanding of life on Earth. By the end of this chapter, you will understand why the future of palaeontology is exciting, with technological advancements and an increased rate of discoveries expected to bring more breakthroughs than ever before.

What is a fossil?

To decipher the history of the Earth, palaeontologists rely on fossils. Derived from the Latin word *fossilis* meaning "unearthed" or "dug up", fossils are naturally preserved remains or impressions of prehistoric life. They are usually over 10,000 years old and found in the Earth's physical structure or substance. Subfossils are anything up to around 50,000 years old, such as giant ground sloths located in South America. The "fossil record" is the entirety of fossils found to date, grouped into two categories:

- Body fossils – parts of organism bodies, such as skeletal remnants, feathers or leaves.
- Trace fossils – the indications of ancient activity, such as footprints, burrows or faeces.

Fossils allow palaeontologists to identify the organisms of the past, but the context in which they are found – or more accurately, preserved – is just as critical to revealing the secrets of ancient life.

Making fossils

Fossilization – the process by which fossils are formed – is rare. Most organisms quickly decompose, but a few get covered by sediments soon after death (usually under ancient seas, lakes and rivers) and begin one of the following processes:

- Petrification/Permineralization – the most common method: mineral-rich water enters the organism and the remains turn to stone.
- Compression – high pressure creates a dark imprint of the remains.
- Moulds and Casts – deposits of sediment or minerals enter the cavities of the organism, causing a three-dimensional impression.
- Replacement – remains dissolve and are replaced by a different mineral.
- Preserved Remains – the rarest method, the organism is frozen, dried or encased by naturally occurring phenomenon including cold storage, natural paraffin and amber.

Fossil hunting

Fossils can lie hidden for thousands, even millions of years. But they are not mined out of the earth like jewels. In fact, most fossils become exposed by the natural processes of erosion. Geological maps are used to locate rocks that are the correct age for the desired fossils. For instance, if you are searching for dinosaur fossils, you need rocks from the Mesozoic Era. Palaeontologists often target deserts because they have sedimentary rocks and no plants or soil covering them up.

Teams of palaeontologists explore huge areas on foot to find fragments of fossils and promising sites. Once a spot is identified, the excavation work begins – with the aid of rock hammers, chisels, pickaxes, shovels and other equipment.

Fossil hunting requires a mix of skill, science, patience and luck. There are also many examples of accidental discoveries. In 1999, the palaeontologist Jason Pool was on a dig in Montana when he went on a toilet break in nearby bushes and stumbled across a rare *Allosaurus* fossil. It became known as *Urinator montanus*.

And if you are not a palaeontologist, you can still contribute to the science. In 2022, nine-year-old fossil collector Molly Sampson picked up a megalodon (a sort of prehistoric shark) tooth on a beach in Maryland – it was 12.7 centimetres long and considered a "once-in-a-lifetime" find. That same year, a designer named Ou Hongtao was at a restaurant in Sichuan province, China, when he spotted oval pits in the stone floor – they were the footprints of two dinosaurs from 100 million years ago.

THE BONE WARS

In the late 1800s, there was a raging rivalry between the American palaeontologists Othniel Charles Marsh and Edward Drinker Cope. In an effort to gain fame and glory, they tried to outdo each other in their quest to find new species of prehistoric creatures. Their use of bribes, tricks and even dynamite scandalized the scientific community and eventually left both men destitute. But their efforts were not in vain; 20 years of competitive fossil hunting led to the discovery of 136 new species of dinosaurs and the first complete skeletons.

How are fossils dated?

Figuring out the age of a fossil is key to understanding it, but the answer is not found in the fossil itself. As the palaeontologist Henry Gee once wrote, "No fossil is buried with its birth certificate". Instead, scientists use two techniques to decipher the age of rocks and fossils, known as absolute and relative dating.

ABSOLUTE (OR NUMERICAL) DATING

Absolute dating tells us the actual age or date range of fossil material and is useful for volcanic rocks and minerals. The most common method is radiometric dating, which analyzes the amount of radioactive decay in the minerals of the rocks. The level of this decay is constant over time and provides the age of any fossils found within the rocks.

RELATIVE DATING

A fossil's relative age can be determined by looking at its position in a layered sequence of sedimentary rocks, a process known as stratigraphy (the flat, horizontal layers of rock are called "strata"). One rock layer, and the fossils within it, will be older or younger than another layer based on its relative position. The "principle of superposition" means that younger rock layers, or strata, are usually above older rock layers. This method puts geological events in chronological order, without a specific numerical age assigned to the rocks.

An alternative method is "rock magnetism" (or palaeomagnetism), which examines the magnetic minerals in ancient rocks to determine the orientation of the Earth's magnetic field at the time the rocks were formed.

The most effective approach for fossil dating is to combine techniques by establishing relative age relationships between local rock units, indexing fossil ages for the sedimentary rocks and radiometric and magnetic dates where possible.

Identifying fossils

Once a fossil is found, palaeontologists begin trying to ascertain what it is. This is complicated because many fossils are poorly preserved, broken or partially covered in the matrix of the surrounding rock. Most fossils are identified based on their shape and forms of symmetry, and then compared and matched with remains from better-preserved specimens. Minute and chemical details can also be analyzed for precise classification.

THE START OF LIFE

Stromatolites are the oldest fossils ever found, dated between 3.5 and 4 billion years old. Curiously, they resemble giant cauliflowers. For 2 billion years, these layered rocks dominated our planet, and the *Cyanobacteria* – a form of microscopic single-celled organism – that formed them, released the oxygen necessary for other life forms to evolve. They are the reason you are alive today! You can still find them in the Bahamas and in Shark Bay, Western Australia.

Beyond hammers and chisels

Picture a palaeontologist and you may envisage someone in a desert chipping at bones in the ground. While fieldwork is integral to the discipline, another site where discoveries occur is the laboratory. Fossils are partial clues about ancient existence, and deciphering them requires the use of cutting-edge technology. In 1969, palaeontologists conducted one of the first computer simulations of animal movement on fossilized trails. In the twenty-first century, technology has triggered what many call the "Golden Age of Palaeontology". For instance:

- Artificial intelligence – scans and analyzes landscapes looking for prime fossil sites.
- Micro-computed tomography scanning (µCT) – a non-destructive tool that creates detailed three-dimensional images to see inside objects.
- Big-data processing – allows specimens from across the world to be compared.

The greatest story ever told

(BYA = billion years ago, MYA = million years ago)

TIME	EVENT
HADEAN EON 4.6 BYA– 4 BYA	Solar system including the Earth forms when meteorites collide and stick to one another, gradually forming a planet (oldest known Earth mineral dated 4.4 billion years old, and several meteorites dated 4.6 billion years old).
ARCHEAN EON 4–2.5 BYA	3.8 BYA Earth is coated with an ocean of shallow salty seas. Single-celled microbes begin to flourish (oldest known fossils are bacteria).
PROTERO-ZOIC EON 2.5 BYA– 541 MYA	Photosynthetic bacteria forms release oxygen into the atmosphere. Many parts of Earth covered in glaciers.
1–2 BYA	First multi-cellular organisms appear on Earth. Life forms gradually evolve and become more complex.

**PHANERO-
ZOIC EON
541 MYA–
PRESENT**

Earth begins to warm, allowing plants and animals to spread across large parts of land.

**541–252 MYA
Paleozoic Era.**

Diversification of visible life beginning with the Cambrian Explosion (sudden appearance of many new complex multicellular organisms in the fossil record). Trilobites and first vertebrates develop (fish with teeth and jaws). Ancient world ocean filled with invertebrate animals including sponges, corals, and crabs.

440 MYA

Land starts to be populated by plants.

370 MYA

Ecosystems develop. Earliest amphibians evolve and the first land vertebrates appear (amphibians).

299 MYA

Enormous continent of Pangea forms.

252 MYA

The Great Dying event: 70 per cent of land vertebrates and 96 per cent of marine species disappear from fossil record (the cause is still debated).

**252 MYA–
65 MYA
Mesozoic Era**

The age of reptiles. Dinosaurs, flying pterosaurs and swimming ichthyosaurs are the dominant vertebrate animals on Earth for nearly 200 million years (during the Triassic, Jurassic and Cretaceous Periods). Land covered with non-flowering gymnosperm plants, the ancestors of modern-day conifer trees (commonly known as evergreen or pine trees). Era ends with the Cretaceous-Paleogene Extinction Event.

**66 MYA–
PRESENT
Cenozoic Era**

The age of mammals. With the dinosaurs gone, mammals begin to fill the roles of large herbivores and carnivores. Modern life forms develop and plants flourish.

55 MYA

First primitive primates emerge (ancestors of monkeys, apes and humans).

35 MYA

Forests decline and grasslands begin to dominate the land.

**2.6 MYA–
11,700 YA**

Global cooling period, known as the "ice age". Many large mammals called megafauna appear, including mammoths, saber-toothed cats, giant ground sloths and mastodons.

0.2 MYA

Anatomically modern humans (*Homo sapiens*) appear.

VERTEBRATES

There is a subdivision of palaeontology that regularly inspires both news headlines and movies. Vertebrate palaeontology involves the study of extinct animals with backbones (or notochords), including fish, amphibians, reptiles, birds, mammals and – you guessed it – dinosaurs. While vertebrates only make up less than 5 per cent of all animal species on our planet, their larger body size and sophisticated adaptations have allowed them to dominate the land and seas for millions of years.

This chapter outlines some of the major vertebrates that have been discovered and explains how scientists connect the animals of the past with their modern-day relatives and descendants. By expanding our knowledge of these beasts, you will see how palaeontologists are uncovering the roots of our own evolution as well as the possibilities that lie ahead for all vertebrates, including ourselves.

Georges Cuvier

In 1812, Georges Cuvier, today known as the "father of palaeontology", told the world it was unlikely any large fossil animal remained undiscovered. While he was to be proven wrong (as the rest of this chapter will show), Cuvier's work was instrumental in the development of vertebrate palaeontology as a discipline. He was the first to compare the skeletons of contemporary living animals with fossils. His work showed that much could be deduced about the structure of fossilized creatures by noting similarities with the bones of contemporary creatures.

In 1796, he presented a paper on living and fossilized elephants. In the face of scepticism, he argued that mammoth and mastodon fossils were distinct from living Indian and African elephants. In doing so, he proved that the mammoth and mastodon fossils belonged to extinct species. Cuvier established the truth of extinction and showing the importance of comparative anatomy.

First descriptions

Megalosaurus is thought to be the first dinosaur described in scientific literature. In 1677, Robert Plot published a written description of the lower part of a massive femur found in an Oxfordshire quarry. Alongside the write-up was the first published illustration of a dinosaur bone. However, he believed the bone was that of a giant human and labelled it *Scrotum humanum*. Eventually, in 1824, William Buckland determined that the fossilized teeth, limb bones and jaws belonged to an extinct giant reptile, which he named *Megalosaurus*.

KIRKDALE CAVE

In 1821, a cave in Yorkshire, England was found filled with the fossilized bones of ancient tigers, hippos, elephants and numerous hyaenas. William Buckland argued that the bones had not drifted there from Africa by a biblical flood, but that the cave was a den for an extinct species of hyaena. Buckland's findings helped to develop ideas about the age of the Earth.

The girl who changed science

Some of the first known fossil discoveries of ancient animals were made by a girl living in Lyme Regis, off the coast of England. In 1811, before the term "dinosaur" even existed, 12-year-old Mary Anning excavated a complete 17 feet (5.2 metres)-long skeleton of a dolphin-like reptile with large eyes. The townspeople at the time believed it to be a monster and scientists assumed it was a crocodile. Following years of study and debate, palaeontologists identified it as 200 million years old. They named it *Ichthyosaurus*.

As Anning grew older and taught herself geology and anatomy, she went on to uncover the sensational remains of many large vertebrates. In 1823, she found the first intact fossil of a *plesiosaur*, the "sea-dragon" that replaced *ichthyosaurs* as the sea's top predators in the Jurassic Period. In 1828, she revealed a "flying-dragon" called *Pterodactylus*. We will never know the full extent of Anning's work, as her gender and lack of social status meant other male and more privileged scientists and collectors took the credit for her findings.

Nevertheless, Anning gained a reputation across Europe and people flocked to Lyme Regis to buy her fossils. She is even said to have inspired the tongue-twister "She sells seashells by the seashore"! Anning's findings shocked the public and academia alike because they were a direct challenge to how the natural world was thought to have developed. The common belief at the time was that everything currently on Earth had always existed since the dawn of time, yet the species revealed by Anning were unlike anything seen before. Today, she is recognized as having laid the groundwork for the theory of evolution. Anning's legacy remains as "the greatest fossilist the world has ever known".

PALAEOART

In 1830, the geologist Henry De la Beche painted a watercolour based on the fossils found by Anning. Titled *Duria Antiquior* or *A More Ancient Dorset*, the image pushed the boundaries of science and art by showing prehistoric animals in their ancient environment. It was the first depiction of dinosaurs as living organisms, and inspired a new art form known as palaeoart.

Terrible lizards

For centuries, people had been encountering fossils that would now be considered dinosaurs. They remained an unsolved mystery until three species of extinct reptiles came to the attention of some pioneering British scientists:

- 1822 – Gideon Mantell and his wife Mary discover the giant fossil teeth of a large reptile that resemble iguanas' teeth. He calls the herbivorous dinosaur *Iguanodon* and estimates the animal died about 130 million years ago.

- 1824 – William Buckland finds fossils of the carnivorous *Megalosaurus*.

- 1831 – Mantell publishes a paper titled *The Age of Reptiles* summarizing evidence of an extended period during which large reptiles dominated the planet.

- 1832 – Mantell finds the partial skeleton of the armoured *Hylaeosaurus*.

These breakthroughs culminated in the trailblazing anatomist Richard Owen declaring in 1841 that a group of extinct Mesozoic reptiles had been discovered. He realized that the *Iguanodon*, *Megalosaurus* and *Hylaeosaurus* shared characteristics that were unlike any other flying or marine reptiles. He gave them the collective term "Dinosauria", meaning "terrible lizards". He also suggested that dinosaurs may have been warm-blooded, like mammals and birds, rather than cold-blooded like other reptiles.

THE START OF DINOMANIA

In 1854, the world's first life-sized sculptures of prehistoric animals, including four dinosaurs (two *Iguanodons*, a *Hylaeosaurus* and a *Megalosaurus*) were revealed at Crystal Palace Park, London. Made by Benjamin Waterhouse Hawkins under the guidance of Richard Owen, the models are famous for showing the scientific inaccuracies of early palaeontology. They did however bring dinosaurs to the attention of the wider public.

Mega shark

Megalodons were gigantic prehistoric fish that lived between 2.6 and 14 million years ago. They might have been the deadliest predators ever. They probably grew to between 15 and 18 metres in length – three times longer than a great white shark. Megalodons' mouths opened to 3.4 metres wide and had 276 teeth. However, no complete skeleton has ever been found. They are known almost entirely by the discovery of their teeth, with the first tooth discovered in 1835 by Swiss-born palaeontologist Louis Agassiz.

THE STRONGEST BITE

Humans can bite down with a force of around 1,317 Newtons (N), while great white sharks have a bite force of approximately 18,216 N. Researchers have estimated that megalodons had a bite of between 108,514 and 182,201 N – more than three times the force of *Tyrannosaurus rex*.

Armoured fish

In the 1830s, Louis Agassiz inspected some fossils of bony armoured fish originating from Scotland. The fish did not resemble any known living creature; they were jawless and had heavy-duty armour on their heads. In 1844, Agassiz classified them as a new group: ostracoderms (meaning shell-skinned). Dating back over 500 million years, they dominated the world's oceans for almost 100 million years.

Agassiz also studied the fossils of other armoured fish called placoderms, which were the first vertebrates to develop jaws, pelvic fins and teeth over 400 million years ago. These jawed fish represented an evolutionary leap toward the skeletal structure present in most animals with a backbone today – including humans.

Giant sloths

The creatures most often associated with palaeontology are dinosaurs, but this was not always so. At the turn of the nineteenth century, mammal fossils were popular, with the giant ground sloth claiming the limelight. In 1787, the first fossilized skeleton was discovered in Argentina and named *Megatherium americanum* (meaning "great beast from America"). These were the largest sloths of all time, standing at over 3.5 metres tall. Since then, over 100 species have been revealed, with the most famous discovery made by Charles Darwin in 1832 and named *Mylodon darwinii*.

DO YOU LOVE AVOCADOS?

Giant ground sloths were one of the few ancient herbivores able to swallow avocados whole. This made them important seed dispersers for these fruits as they could hold them in their digestive tracts and eventually defecate them far away from the parent tree.

Bipedal lizards

Richard Owen's theory that the Earth was once inhabited by gargantuan creatures was wildly controversial in 1841. In 1858, the nearly complete skeleton of a gigantic prehistoric animal was excavated in Haddonfield in the U.S. by William Parker Foulke. Luckily, Foulke was friends with Dr Joseph Leidy, the founder of vertebrate palaeontology in the U.S., and he invited him to the site. Dr Leidy recognized that the 7.62-metre-long creature, which had anatomical features of both a lizard and a bird, was a dinosaur like the recently discovered *Iguanodon.* He named it *Hadrosaurus foulkii* (meaning heavy or bulky lizard). The bones proved the anatomy of dinosaurs had been as different from modern animals as Owens had predicted. Based on the orientation of the pelvis, Leidy concluded that this new specimen was bipedal with an upright posture. Bipedalism transformed the understanding of dinosaur appearance and posture.

The king lizard

The *Tyrannosaurus rex* needs little introduction, but you may not be aware of another behemoth that lived in the oceans. *Basilosaurus*, or the "king lizard", had an unusually long body and superficially reptile-like skin. The first complete skeleton was discovered in 1834 and so-called by Richard Harlan. Later studies revealed it was not a lizard but a prehistoric whale that dominated the seas 40–34 million years ago – after dinosaurs had gone extinct. They were the top predators of their environment, preying on sharks, large fish and other marine mammals.

FOSSIL FURNITURE

In the nineteenth century, *Basilosaurus* fossils were so common in the southern parts of the U.S. that their large vertebrae were often used for furniture and building construction materials. It was designated the Alabama state fossil in 1984.

Everybody poops

A less glamorous yet highly transformative find in the nineteenth century was of the coprolite. Coprolites are the fossilized faeces of ancient animals. They were first identified by Mary Anning and William Buckland in the 1820s. By looking at the shape and size of coprolites, palaeontologists can establish what kind of animal produced them and where they lived. Coprolites can also contain clues about an animal's diet and ancient food chains. In 2005, 65-million-year-old coprolites proved some dinosaurs dined on grass – a surprise when grasses were previously thought to have evolved after dinosaurs went extinct. Some coprolites bear traces of preserved muscle, allowing scientists to calculate how quickly an organism's digestive system broke down food.

The transitional fossil

In 1859, the scientific world was scandalized by the theory of natural selection. The British naturalists Alfred Russel Wallace and Charles Darwin argued species formed and changed through natural selection – or "the survival of the fittest", to use the philosopher Herbert Spencer's later term – without divine intervention. But one question still baffled everyone: how did the world become repopulated with animals after mass extinction events? The answer came two years later with the discovery of *Archaeopteryx*, the first reptilian fossil found with feathers, wings and a long bony tail. At 150 million years old, it provided evidence of a species transition – from dinosaurs to birds.

A SNAPSHOT OF JURASSIC LIFE

The first *Archaeopteryx* fossil was found in the Solnhofen limestone formation in Bavaria, Germany. Since 1861, this limestone has revealed over 750 fossilized organisms, including jellyfish and insects.

Dinosaur diplomacy

When railroad workers unearthed the near-complete fossilized bones of a *Diplodocus* in Wyoming, U.S., in 1899, newspapers billed the discovery as "the most colossal animal ever on Earth". Thought to be between 152 and 154 million years old, the Wyoming discovery became one of the most easily identifiable dinosaurs. The Scottish-born industrialist Andrew Carnegie arranged for 21-metre-long, 4.25-metre-high plaster replicas to be made of some *Diplodocuses* and gifted them to museums around the world.

SIZE MATTERS

Diplodocus fossils have been preserved at different stages of growth. As *Diplodocuses* grew to over 100 feet, their fossils help to demonstrate how they grew to such massive sizes.

King of the dinosaurs

The first *Tyrannosaurus rex* (also known as T-Rex) fossils were unearthed in 1892 by Edward Cope, but he failed to identify them correctly (a fact that was not realized until 2000). Barnum Brown is recognized as the person who first discovered T-Rex fossils in 1902, in Montana, U.S. Since then, several well-preserved fossils have been found. In 1990, a 65-million-year-old skeleton nicknamed Sue showed T-Rex was one of the largest carnivorous dinosaurs to have ever lived.

ALMOST DYNAMOSAURUS

In 1905, two skeletons of the same creature were mentioned in a publication but with different names, *Dynamosaurus imperiosus* (meaning the "Imperial Powerful Lizard") and *Tyrannosaurus rex*. The fossil, referred to as *Dynamosaurus,* was discovered first but in the paper it was the later find, named *Tyrannosaurus rex,* that got *described first* and therefore took precedence.

Tendaguru

In 1906, miners found enormous bones near Tendaguru Hill in German East Africa (what is now Tanzania). German palaeontologists and local labourers uncovered over 250 tonnes of fossils, representing hundreds of species of invertebrates, fish, early mammals and reptiles from the Late Jurassic Period. Among them were the fossils of what would become the biggest mounted dinosaur on Earth: *Brachiosaurus brancai* (later renamed *Giraffatitan brancai*). The area is now called the Tendaguru Palaeontological Site and is considered the richest Late Jurassic strata in Africa.

LAGERSTÄTTE

Lagerstätte means a sedimentary deposit that exhibits well-preserved fossils – sometimes including soft tissues (such as skin, internal organs, blood vessels and nerves). The Tendaguru Formation is one example.

Flying reptiles

Like their cousins the dinosaurs, *Pterosaurs* (wing-lizards) are considered one of evolution's great success stories. The first species was found in Germany and described by Italian historian Cosimo Alessandro Collini in 1784. Georges Cuvier identified it as a winged reptile when he studied the fossils in 1801. He called it "ptero-dactyle" (a name since revised to *Pterodactylus*). *Pterosaur* fossils were then discovered in North America in 1871. These specimens demonstrated that they inhabited different continents and were the first vertebrates to evolve powered flight about 220 million years ago. As the discoveries increased, palaeontologists recognized that *Pterosaurs* lived throughout the Mesozoic Era. By the Late Cretaceous Period, their wingspan ranged from 2 to 11 metres. This made them the largest known flying animals.

The mystery was finally solved in 2020 when a 237-million-year-old reptile group called *lagerpetids* was acknowledged as the evolutionary precursor to *Pterosaurs*. *Lagerpetids* were only known by a few partial skeletons (found in the U.S., Argentina, Brazil and Madagascar) but scanning technology allowed palaeontologists to identify common skeletal traits between the two species.

A SURGE IN UNDERSTANDING

In 2014, 200 eggs and 16 embryos from the *Pterosaur Hamipterus* were discovered in China. In 2017, the largest ever skeleton of a *Pterosaur* was discovered by Amelia Penny on the Isle of Skye, Scotland. The new species was named *Dearc sgiathanach* (pronounced jark ski-an-ach), meaning "winged reptile".

Giant wombat

Diprotodon, meaning "two forward teeth", is the largest known marsupial (pouched mammal) ever to have existed – standing at up to 1.8 metres tall, and weighing nearly 3,000 kilograms. It had the appearance of a wombat but the proportions of a rhinoceros. *Diprotodon* roamed the Earth for around 2.5 million years before it went extinct about 55,000 years ago.

EXTINCT MEGAFAUNA

Megafauna are any animals with an adult body weight of over 45 kilograms. Today, this includes elephants, giraffes, cows, deer and humans. Before there were widespread human settlements (resulting in the killing of the biggest beasts), animals were free to grow to mind-boggling proportions. Examples include the *argentavis* (a 7.3-metre-long bird) and *megalania* (a lizard that grew up to 7 metres).

In eternal combat

In 1971, a fossil of a *Velociraptor* and a *Protoceratops* locked in combat was uncovered in the Gobi desert, Mongolia. The fossil shows the *Velociraptor* had embedded its foot claw into the neck of the crouching *Protoceratops*. In turn, the *Protoceratops* had bitten the right arm of the *Velociraptor*. The discovery showed that non-avian theropods (carnivorous dinosaurs with three toes and claws) could be active predators rather than mere scavengers.

THE DUELLING DINOSAURS

The fossilized skeletons of a T-Rex and *Triceratops* intertwined in a fight was discovered on a ranch in Montana, U.S., in 2006. They are among the most complete skeletons ever discovered of the two iconic dinosaurs. Legal battles over their ownership delayed their public display until 2023.

The first tetrapods

"It's every palaeontologist's dream to find a transitional form, something that falls between two groups that we are familiar with…"

Jenny Clack

How did vertebrates make the transition from sea to land? For decades it was assumed that our distant ancestors came ashore and then evolved legs. In 1987, British palaeontologist Jenny Clack discovered in Greenland, a 360-million-year-old specimen of a fish-like animal with limbs that was later named *Acanthostega,* along with other early *Tetrapods* (animals with four legs) such as the *Ichthyostega.* These fossils indicated that limbs evolved over tens of millions of years while the creatures were still water-bound. These fossils marked one of the biggest transitions in evolution.

Trace fossils

Trace fossils, such as footprints, are incredibly rare and hard to spot. The "Valentia *Tetrapod* Trackway" was discovered in 1993 on Valentia Island, Ireland; its impressions made by early land animals about 383–388 million years ago. There are at least five separate tracks, the longest measuring 15 metres and comprising over 145 footprints of an early amphibian. Some of the shorter trackways show traces of a tail having dragged along the silty mud. These trace fossils represent a turning point in evolution when creatures first started to walk on land.

RUNNING IN THE RAIN

A rare set of dinosaur footprints discovered near Jinju City, South Korea, in 2019 contained well-preserved skin surrounded by water drop impressions. The tracks, left behind by a *Minisauripus*, were probably made during or just after a rain shower about 112 million years ago.

Preparing for land

In 2004, the story of how vertebrates transitioned onto land was elaborated by a 375-million-year-old fish called *Tiktaalik*. The fossilized remains were found on Ellesmere Island in the Canadian Arctic and were a blend of gills, fins, scales, lungs, a mobile neck, a ribcage and a flattened crocodile-like head. Scientists were surprised to find that this 3-metre-long creature had a large pelvis and powerful hind fins despite living in water and not yet adapting to life on land. These limbs could have propelled the fish through water, but also helped it walk on riverbeds or scramble around mudflats.

OLDEST FOOTPRINTS ON EARTH

Polish palaeontologist Grzegorz Niedźwiedzki discovered 395-million-year-old traces of footprints in 2010. The fossil shows the tracks of a land vertebrate plodding across a mud-flat, with clear traces of toes.

The Titanosaur

One of the largest creatures ever was excavated in Patagonia, Argentina, from 2012 to 2015. The most complete giant dinosaur ever discovered, this 57-tonne behemoth was known as the *Titanosaur* until it was named *Patagotitan mayorum* in 2017. It was an estimated 37 metres from nose to tail. We are still not sure why and how some dinosaurs grew to such proportions.

SAUROPODS AND TITANOSAURS

Sauropods are long-necked herbivorous dinosaurs like *Brontosaurus*, *Diplodocus* and *Brachiosaurus*. They are thought to have lived all over the world, based on fossil discoveries in the Antarctic. A subgroup called the *Titanosauria* contained the largest *Sauropods* including *Dreadnoughtus* and *Argentinosaurus*.

Nests and eggs

In 2011 in Mongolia, palaeontologist David Fastovsky unearthed a 70-million-year-old nest containing the remains of 15 baby dinosaurs of the species *Protoceratops andrewsi*. This was the first proof that young dinosaurs stayed in the nest for an extended period. Chemical analyses revealed that all 15 dinosaurs were roughly the same size and at the same developmental stage, which suggested that they shared the same mother. This helped substantiate the claim that some dinosaurs, including *Protoceratops*, cared for their young during early childhood.

FOSSILIZED EGGS

In 2023, palaeontologists in India located a fossilized dinosaur hatchery of 92 nests and 256 eggs belonging to giant plant-eating *titanosaurs*. The eggs ranged between 15 and 17 centimetres in diameter. The closely spaced nests suggest *titanosaurs* laid their eggs and left the babies to fend for themselves.

Buried birth

Two hundred and forty-eight million years ago, an *ichthyosaur* mother was giving birth when it died and eventually fossilized. The remains were discovered in China in 2011, with one baby already born outside the body, a second still inside waiting to be born and a third trapped in the birth canal. This fossil shattered the belief that *ichthyosaurs* gave birth in water, not on land. The fossilized offspring were emerging head-first – a behaviour found only in animals that give birth on land.

SWITCHING TO LIVE BIRTHS

Generally, reptiles lay eggs and mammals give birth to live young. But in 2013, researchers revealed that reptiles historically veered back and forth between the two strategies (in response to ecological conditions) before settling on egg-laying. Colder temperatures seemed to prompt the switch to live births because they enable a female to retain her eggs for longer periods.

The sleeping dragon

In 2011, one of the most incredible discoveries was unearthed: the "sleeping *Nodosaur*". The 112-million-year-old fossil was a new species of armoured dinosaur named *Borealopelta markmitchelli* (a member of the *Nodosauridae* family of dinosaurs). It likely weighed around 1,300 kilograms and measured 5.5 metres. Its three-dimensional preservation gives the appearance that it is merely sleeping. It offers one of the most lifelike images of a dinosaur ever found. Even its stomach contents were preserved, which is unusual with herbivores; using a microscope, scientists could see the cellular detail of the plants it had eaten. Palaeontologists ascertained the colour of the armoured beast using chemical analysis, which revealed a reddish-brown pigment. This indicated that camouflage was required to protect it from predators.

Not just any brain

In 2023, palaeontologists came across the oldest three-dimensional fossil brain ever discovered. The skull belonged to the extinct *Coccocephalus wildi* and was found in a coal mine in England more than a century prior to the scan. It had been sitting undiscovered in the archives of a museum. The palaeontologist Sam Giles said that it filled important gaps in the knowledge of brain evolution in fish, particularly the early evolution of ray-finned fish. There are approximately 30,000 ray-finned fish species, which account for about half of all backboned animal species. Significantly, *Coccocephalus'* brain was established to have folded inward, unlike living ray-finned fish in which the brain folds outward. This provided an insight into when the trait originated.

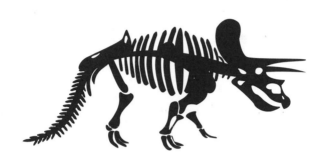

INVERTEBRATES

Invertebrate palaeontology is the study of fossil animals that lack a backbone or bony skeleton. They range in size from microscopic plankton and worms to 700-kilogram squids that have the largest eyes of any creature known to have existed. Familiar invertebrates include insects like butterflies, spiders and crabs. Ninety-seven per cent of animal species on Earth are invertebrates, and about three-quarters of these are insects! This abundance was also true in the past – most animal fossil discoveries are invertebrates spanning over 600 million years. Their historic remains usually appear as trackways, shells, casts and moulds, faecal pellets, tubes and exoskeletons (a hard or protective structure outside an animal's body). Their diversity of adaptations reveal how our planet has changed over time.

Small but mighty

Invertebrate fossils are more common and easier to identify than vertebrate fossils because:

- Many invertebrates are small, making them less vulnerable to breakage by the various forces of nature.
- Their small size means they are more likely to remain intact as a complete fossil.
- Many prehistoric invertebrates possessed durable hard parts such as trilobites, clams and corals – these parts are relatively simple compared with the multi-jointed skeletons of vertebrates.
- Hard shelled invertebrates were often buried underwater in sediment-preserving conditions.

While one extraordinary vertebrate fossil can attract attention, a thousand fossils of an invertebrate can reveal a whole chapter in the story of life on Earth.

Different invertebrates

Jean-Baptiste de Lamarck, a pioneering biologist and evolutionist, first conceptualized and coined the category "Invertebrata" between 1793 and 1801. When it comes to the fossil record, invertebrates have been further divided into the following categories:

- Soft-bodied and miniscule invertebrates that are rarely fossilized. Examples include jellies, flatworms and nematodes.

- Hard-bodied and large invertebrates that are more commonly preserved due to their hard parts. Examples include creatures with shells, plates, jaws and teeth.

Soft-bodied invertebrates are usually identified by trace fossils, microfossils (tiny remains seen through a microscope) or chemofossil residues (chemicals left behind by dead biomaterial).

The foundations

*"Invertebrate animals… show us, much better than
the higher animals, the true course of nature."*

Jean-Baptiste de Lamarck

In the 1500s, the German scholar Georgius Agricola
attempted to categorize minerals, rocks and sediments
in his text *De Natura Fossilium*. In the 1700s, French
geologist Jean-Étienne Guettard studied ancient molluscs
and siliceous sponges while identifying several fossil
species. But it was not until 1793 that the study and
classification of invertebrates truly began. This was the
year Jean-Baptiste de Lamarck was appointed Professor
of Insects and Worms at the National Museum of
Natural History in Paris. He went on to write his seminal
publication *Mémoires sur les Fossiles des Environs de Paris*
which laid the foundations of invertebrate palaeontology.
Lamarck also presented a theory of evolution based on
his transformative idea that complex organisms had
evolved from simpler ones over time.

Natural spirals

Ammonites are molluscs that lived in the oceans between 400 and 650 million years ago, and are closely related to squid, octopuses and cuttlefish. Characterized by ribbed spiral-form shells, they evolved in myriad shapes and sizes. Ammonites were so prolific that over 10,000 species have been discovered. Because they were widespread and evolved rapidly, they help to date the rocks they were fossilized in, as well as determine the age of other fossilized creatures located in the same rocks.

One example of their abundant numbers is the "Marston Magna Marble", found in the small village of Marston Magna in Somerset, England. It is not a true marble but rather a 200-million-year-old piece of limestone that is crowded with small white fossilized ammonites. Most known specimens were collected during the eighteenth and nineteenth centuries, when some of the slabs were used to make grave headstones.

The trigger for diversity

In 1689, Reverend Edward Lhwyd described "some kind of flat fish" collected in Wales. This was the first report of fossilized invertebrates known as trilobites, of which 17,000 species have been identified. These advanced marine creatures lived between 250 and 520 million years ago and are closely related to crabs, spiders and lobsters. Like today's horseshoe crabs, they likely scuttled around the ocean floor searching for food – and developed eyes to help them. They are the oldest fossils with complex eyes that included numerous lenses (up to 15,000!) for focusing. Palaeontologists believe that trilobite eyes may have compelled other marine organisms to evolve defence mechanisms like propulsion, camouflage and sight – thereby triggering the explosion of diverse life forms that led to the dinosaurs.

Burgess Shale

In the late summer of 1909, the palaeontologist Charles Walcott was walking in the Canadian Rockies with his wife and their horse. Reportedly, the horse slipped and overturned some rocks that turned out to be 509 million years old. The rocks came to be known as the Burgess Shale and contain arguably the most important animal fossils ever found.

The fossils totalled 65,000 specimens representing about 127 species. Dating from the Cambrian Explosion Era, the creatures existed at the very beginning of complex life on Earth. Unlike the previous 3 billion years, the oceans were suddenly teeming with new life forms that were creeping, burrowing and swimming. Most fossils from this period are remains of shells and other hard parts, but the Burgess Shale contained entire organisms with eyes and soft tissue preserved.

While some of the fossilized creatures were known, such as trilobites, many were completely novel and odd. For example:

- *Anomalocaris* – measuring up to 3 feet long, this predator had a pair of large eyes at the front and a rounded mouth with sharp teeth.

- *Aysheaia* – around 5 centimetres in length, it had ten pairs of stubby little legs that bore small spikes.

- *Hallucigenia* – a centimetre-long worm that looked like it was walking on stilts.

- *Marrella splendens* – a 25-millimetre-long lace crab that breathed by kicking its legs. Each body segment had a pair of legs, and each leg had a gill.

- *Opabinia* – a creature with five eyes arranged on stalks, a long slender body and claw-like appendages. When its image was first presented at a scientific conference in 1972, people laughed as they thought it was a joke!

- *Pikaia* – a 3.8-centimetre creature with a spinal rod called a notochord, the earliest known chordate – the group of animals that would later evolve into vertebrates.

Gunflint chert

Palaeontological studies of ancient invertebrates were sporadic and often discredited until the work of Stanley Tyler and Elso Barghoorn in the 1950s. Their reports on microfossils found in the Gunflint Iron Formation (Gunflint chert) in Ontario, Canada, provided the first detailed, irrefutable evidence for life in the Precambrian Era. At the time of is excavation, the rock layer contained the earliest form of life discovered and described in scientific literature, as well as the earliest evidence for photosynthetic organisms. The diverse and structurally preserved fossils of microbes were dated between 1.9 and 2.3 billion years old. This meant the atmosphere was oxygenated enough by this time to sustain microbial life. Tyler and Barghoorn's findings that life occurred during the Precambrian were published in 1965 and augured an academic rush to explore Precambrian microfossils elsewhere. Though the Gunflint chert no longer represents the oldest life discovered on Earth, it proved that photosynthesis and the dawn of life started a billion years earlier than previous estimates.

Primitive dragonflies

Some of the first winged insects in the world were *Protodonata* that lived between 290 and 360 million years ago. These creatures were early ancestors of modern dragonflies, with long narrow bodies, huge eyes, strong jaws and spiny legs for grasping prey. However, the structure of their wings was different – they were primitive in the number and pattern of their veins, and much bigger. Their wingspan could reach up to 75 centimetres and they weighed over 450 grams, similar to crows. They lived when oxygen levels were higher than today (35 per cent of the atmosphere instead of today's 21 per cent), which may have allowed gigantic insects to evolve. The largest species was called *Meganeura*. The fossils are mostly just fragments of wings, although a few full wings and body impressions have been found. In 1978, two miners dislodged a fossil of a 300-million-year-old *Meganeura* in Bolsover Colliery, England. The giant insect had a wingspan of half a metre and was dubbed the "Bolsover dragonfly".

Where did life begin?

Throughout Earth's history, organisms have adapted to harsh conditions. One example is creatures that live within hydrothermal vents. Hydrothermal vents are one of the earliest types of environment to have existed on Earth, having been a feature of the world's oceans since the Hadean Eon (the first geological era after Earth's formation), 4–4.6 billion years ago. They were first uncovered in 1977 when scientists were exploring the seabed in the Galápagos Rift region of the East Pacific. The fluid released from the vents could reach over 400 degrees Celsius, yet an entire living community surrounded them, powered by geothermal energy rather than sunlight. This is because the fluid cools quickly so that temperatures near to the vents are closer to 20 degrees Celsius. The food chain relies on a process called chemosynthesis, which is carried out by bacteria. Like photosynthesis used by plants on land, the bacteria use the energy in chemicals drawn from the vent fluid to convert carbon dioxide into food.

THE LOST CITY

In 2000, a field of alkaline hydrothermal vents
was found in the middle of the Atlantic Ocean
filled with trillions of microbial organisms found
nowhere else on Earth. Known as the "Lost City",
it is estimated to be over 120,000 years old.

Some geologists and biologists believe that these ocean floor oases are the sites where life on our planet began. Fossils from hydrothermal vents are among the earliest evidence for Earth's fauna, with a range of specimens showing that vent habitats were readily exploited by microbes during the Precambrian.

The oldest direct evidence of life on Earth was found in Eastern Canada in 2017. Known as the "Nuvvuagittuq" fossils, they show the remains of microbes that lived 3.8 to 4.3 billion years ago (Earth is understood to have formed 4.6 billion years ago). The microbes were formed by bacteria that obtained energy by oxidizing iron minerals in a system of deep-sea hydrothermal vents.

The first animal embryos?

In the late 1990s, the Doushantuo Formation was discovered in Southern China. The 560- to 580-million-year-old soft-bodied fossils resembled the embryos of modern animals. The fossils predated the Cambrian Explosion and were so beautifully preserved that cellular structures were visible under magnification, including soft tissues. Most shocking of all was the evidence of bilateral symmetry (see page 70), which is a key characteristic in many modern animals and believed to have evolved later, during the Cambrian Explosion. Unlike the rigid membranes of algae, they were also shaped by adjoining cells, implying a flexible membrane. Some palaeontologists have suggested that they are not embryos (or bacteria, algae or fungi) but a form of parasite that could multiply its cells and explode them as "spores" to breed the next generation. While the argument continues about what these organisms really are, the Doushantuo fossils offer evidence that major evolutionary diversification of animals had started before the Cambrian Period.

Collective behaviour

Collective behaviour is rare in modern-day invertebrates, but it may have occurred early in evolution. In 2008, researchers were amazed to discover 525-million-year-old fossils of tiny, shrimp-like invertebrates linked together like daisy chains of up to 20 individuals. The tail of each animal was inserted into the exoskeleton of the next. Further evidence was unearthed in 2019 near the town of Zagora, Morocco. A 480-million-year-old fossil revealed a dozen trilobites in a row all facing in the same direction. Palaeontologists have suggested possible reasons why creatures marched in single file:

- To avoid bad weather, moving from one micro-environment to another (like today's spiny lobsters).
- Seasonal reproductive behaviour, such as the migration of sexually mature individuals to spawning grounds.
- A defence mechanism against predators.

Death marches

It is incredibly rare to find trackways and trace makers preserved together in the fossil record (known as a mortichnium). In 2012, palaeontologists discovered the longest complete fossilized death track ever found, measuring 9.7 metres. The fossil was located in Solnhofen, Germany, an area famous for its excellent preservation of fossils. The trackway was dated to around 150 million years ago and displayed the footprints of a horseshoe crab, telson (tail) drag marks, a few head prints and several walking styles. The specimen helped palaeontologists understand what was happening to the animal in the final moments of its life.

In 2019, the fossil of a worm that lived 550 million years ago was found in central China, together with its final tracks. Named *Yilingia spiciformis*, the fossil represented the earliest known independent movement of segmented animals, and the first signs of decision making among animals (by moving toward or away from objects on the ocean floor).

Roots of the insect kingdom

Insects were the first animals on Earth to develop flight, which has been key to their success – there are over 750,000 living species. Scientists believe they were just as prevalent in the past, though the fossil record is patchy. For decades, there were few if any confirmed insect fossils from between 385 and 325 million years ago, a period known as the Hexapoda Gap. In 2012, the paucity of knowledge was finally narrowed by the discovery of a complete fossil in Belgium, known as *Strudiella devonica*. The fossilized insect was 8 millimetres long, 370 million years old and wingless, but its features implied that, by this point, insects had already started to diversify.

ARTHROPODS

Arthropods are invertebrates – they have external skeletons, segmented bodies and paired legs. Arthro means jointed, and pod means foot. Insects are part of the larger arthropod group – they have a three-part body, three pairs of legs, compound eyes and antennae.

Early diversification

In 2012, a 520-million-year-old fossil site located in southwest China, the Chengjiang biota, was declared a UNESCO World Heritage site. Since the first specimens were discovered in 1984, many thousands of fossils have been collected. Like the celebrated Burgess Shale site in Canada, Chengjiang contains a diverse spectrum of marine fossils. The creatures' hard skeletal parts were preserved in exquisite detail (which is typical of most sedimentary deposits), but so were nonmineralized skeletal parts and internal soft parts (which is much more unusual in sedimentary deposits). They provide important scientific information about the origins of animal groups that have sustained global biodiversity to the present day. One of the many incredible species discovered was the *Jianshanopodia decora*, an intermediary creature between Cambrian soft-bodied organisms and today's *Arthropods* (such as insects, millipedes and centipedes). With branching limbs and the ability to swim and crawl on the sea floor, it seemingly sucked up prey using its "trunk" and measured up to 40 centimetres in length.

Sociality

Termite soldiers and ant battles sealed in 100-million-year-old amber were found in Myanmar in 2016. The pieces of amber are some the oldest examples of advanced social behaviour in animals. This led to the insight that ants and termites achieved the adaptation of organizing themselves into hierarchies, which enabled them to survive and flourish to the present day. The diverse anatomy of the termites proved that caste systems were already in place by this time, with reproductive specialization into worker and queen castes. Two worker ants of different species had been fighting before being entombed in the amber.

SPIDER ATTACK

In 2012, a 100-million-year-old amber block was found containing a wasp being attacked by a spider. The moment was so well preserved that 15 strands of spider silk were also found with the duelling minibeasts.

Monster worms!

A new species of gigantic prehistoric worm was discovered in 2017 from jaw fossils that had been sitting ignored in a museum collection since 1994. Known as *Websteroprion armstrongi* (as a tribute to Alex Webster, bass player with death metal band Cannibal Corpse), this species of giant bristle worm had jaws that measured over 1 centimetre in length. Though this sounds small, fossil polychaete jaws (the only hard part in these animals) usually measure just 0.1-2 millimetres, making *Websteroprion*'s jaws colossal in comparison. Based on the jaw fossils, scientists estimate that *Websteroprion armstrongi* could have been 1-2 metres in length. It burrowed deep under the sediment of the ocean floor, with only its jaws above ground waiting for unsuspecting prey. As Sebastian Kvist, curator of invertebrates at the Royal Ontario Museum in Toronto said, "It is a very mean worm". The discovery is another example of ancient gigantism and shows that this ecologically important trait was present in worms 400 million years ago.

Giant millipede

A fossil of a rare giant millipede named *Arthropleura* was discovered in 2018. The specimen, found on a beach in Northumberland, England, was made up of multiple articulated exoskeleton segments, broadly similar in form to modern millipedes. Dating from 326 million years ago, it was the oldest *Arthropleura* fossil ever found, and it set a new world record for the largest land invertebrate to have ever lived. The original creature was estimated to be more than 2.5 metres long and around 50 kilograms in weight. Scientists previously believed that the creatures grew to such massive proportions due to a peak in atmospheric oxygen during the late Carboniferous and Permian Periods. However, this fossil came from rocks deposited before this peak, which showed that oxygen could not be the only explanation. Researchers understood that it must have had a high-nutrient diet, which could have meant it supplemented a diet of nuts and seeds with other creatures, including small vertebrates such as amphibians.

Front and back

What do worms, insects, dinosaurs and humans have in common? They are all animals organized around the same basic bilaterian body plan: a front and back, two symmetrical sides and openings at either end joined by a gut. The development of this bilateral symmetry was a key event in the evolution of life on Earth, but the evidence of its origins was hard to find. Scientists had suspected that 555-million-year-old fossilized burrows found in Nilpena, South Australia, were made by bilaterians, but there was no sign of the creature that made the burrows. Finally, in 2020, a fossilized organism about the size of a grain of rice was discovered in Australia and proved to be the earliest example of a bilaterian. With the help of three-dimensional laser scanning, researchers were able to see a cylindrical body with a distinct head and tail and faintly grooved musculature. The depth and curvature of the creature showed distinct front and rear ends, supporting the directed movement found in the burrows. The creature was named *Ikaria wariootia* and, as predicted, it lived 555 million years ago.

Earth's earliest animal?

Palaeontologists have long debated when complex animals first evolved. Most major groups of animals – including *Arthropods*, molluscs and worms – first appear in the fossil record during the Cambrian Explosion 541 million years ago. The lack of animal fossils before this time could be because older creatures were rarely preserved as fossils (as soft-bodied animals do not usually fossilize well), or complex organisms simply did not exist before then. However, in 2021, Canadian invertebrate palaeontologist Elizabeth Turner reported that animals may have formed around 890 million years ago – 349 million years earlier than previously thought. Turner had collected fossils from ancient microbial reefs preserved in the rocks of Canada's remote Northwest Territories. Using light microscopy technology to examine the rocks, Turner discovered microscopic networks of connecting tubes of calcium carbonate that branched to form irregular three-dimensional shapes – a mesh-like structure similar to modern bath sponges. If the analysis is correct, these would be the oldest animal fossils ever discovered.

Army ants

For almost 100 years, an amber fossil labelled as a common type of ant lay untouched in the archives of the Museum of Comparative Zoology at Harvard University. In 2022, a student named Christine Sosiak happened to put the specimen under a microscope and discovered that it was something different and much more significant: it was the oldest army ant ever discovered, at 35 million years old. It was the first evidence that these ferocious predators once roamed Continental Europe.

NEVER OUTNUMBERED

Army ants are so-called due to their collective foraging skills. Armies up to 15 million strong march across forest floors, consuming other insects but also much larger animals, including lizards, birds and small mammals.

A 500-million-year-old mystery

Many of the animals that appeared on Earth around 540 million years ago consisted of short, simple, hollow tubes. The soft components were rarely fossilized, making it practically impossible to determine which creatures formed these skeletons. In 2017, PhD student Guangxu Zhang found fossils in China's Yunnan province that finally illuminated these earliest skeleton makers. The 514-million-year-old fossils were of a tube-dwelling creature named *Gangtoucunia aspera* and the specimens included soft tissue impressions. Even the stomach and mouth parts were preserved. In 2022, scientists concluded that the tubes belonged to an ancient skeleton-making jellyfish, who were among the first on Earth to develop hard bones.

The ancient octopus

The oldest known ancestor of octopuses – an approximately 330-million-year-old fossil – was revealed in 2022. The fossil was originally excavated in Montana in 1988 and stored in a museum drawer until palaeontologists noticed that the specimen had ten limbs. The finding proved the ancestry of octopuses could be traced back to millions of years before the dinosaur. It also showed how the oldest and earliest diverging relatives of octopuses, the *Vampyropods*, might have looked. The fossil was the first and only known *Vampyropod* with ten functional appendages. The ten arms had suckers attached – unusual since these arms were just muscle. Palaeontologists named it *Syllipsimopodi bideni*, in support of U.S. President Joe Biden's plans to counter climate change.

A ghost lineage

There is a creature that can survive boiling water, freezing water, asteroid hits, mass extinctions and even the vacuum of space. *Tardigrades*, otherwise known as water bears or moss piglets due to their appearance, measure 1 millimetre or less and are the smallest known animals with legs. They look like a cross between a pig and a bear, and have survived on Earth for over 500 million years.

Despite their long evolutionary history, fossils are hard to find due to their microscopic size and non-biomineralizing body. In 2021, the first fossil from the Cenozoic Era was found in a 16-million-year-old piece of amber mined from the Dominican Republic. Scientists were able to use a confocal laser microscope to examine its anatomical features. This yielded information about how the 1,300 *Tardigrade* species that exist today evolved.

Evolutionary riddle

For centuries, there has been a gap in the fossil record that links the evolution of invertebrates to the first vertebrates. In 2022, Baoyu Jiang and his colleagues found that *Yunnanozoans* are the oldest known stem vertebrates. *Yunnanozoans* date back to the early Cambrian Period, about 518 million years ago. The worm-like fish was 3–4 centimetres long, and many of its fossils have been found in Yunnan province, China (hence the name). Jiang and his colleagues analyzed fossil specimens with high-resolution imaging and geochemical methods. The creature's symmetrical arches were found to contain a pattern of filaments, interconnected by bars of cellular cartilage – a feature considered specific to vertebrates.

STEM VERTEBRATE

A stem vertebrate is an extinct vertebrate creature that is very closely related to living vertebrates.

Jumbo shrimp

In 2022, the Taichoute fossil site in Morocco revealed that giant *Arthropods* measuring up to 2 metres long dominated the seas 470 million years ago. The size was a significant discovery given their gigantic proportions in comparison with today's shrimps, insects and spiders that are descendants of these early invertebrates. Even soft parts of the animals, including internal organs, were preserved. The find allowed scientists to investigate the anatomy of early animal life on Earth and how it has changed since.

THE OLDEST PENIS EVER FOUND

The fossil of a 425-million-year-old *ostracod*, or seed shrimp, was uncovered in 2003. It had a disproportionately large penis. The shell and the shape of the creature's soft tissue were preserved in three dimensions.

Ancient sounds

An important evolutionary development in animals was acoustic communication. Insects were the first land animals to send sound waves through the air to communicate over distances. In 2023, the earliest known ears and sound-producing organs were found in fossils of grasshopper-like insects called *Katydids*, dating from 150 to 240 million years ago. They probably evolved calls to avoid detection by predators, helping to drive the co-evolution of hearing in their mammalian predators.

PRECURSOR TO POLLINATION

The oldest known insects covered in pollen – and potentially some of the world's first plant pollinators – were discovered in 2023. Known as *tillyardembiids*, the winged insects were 280 million years old, and the pollen came from a handful of seed-producing, nonflowering plants called *gymnosperms* (see page 103). This suggested early plant–insect interactions were specialized and occurred millions of years before flowers.

PALAEOBOTANY

The fossilized bones of dinosaurs and other giant creatures that once roamed our planet are undoubtedly spectacular and worthy of attention. But what the news frequently ignores are the specimens that often appear alongside the bones: the fossils of ancient plants. This is where palaeobotany comes in; this branch of palaeontology studies terrestrial (land-dwelling) plant fossils as well as prehistoric marine autotrophs (organisms that produced their own light) such as photosynthetic algae, weeds or kelps. This chapter introduces you to some of the incredible discoveries made by palaeobotanists and explains why these breakthroughs offer key insights into Earth's history, such as:

- The earliest occurrence of different kinds of plants.
- Relationships among groups of plants.
- Environments in which plants grew.
- Vegetation responses to past global change.
- Information about the animals that ate the plants.

You will also see why learning about ancient plants can reveal clues about Earth's climatic future.

Organizing fossil plants

People throughout the world have been finding and collecting plant fossils for hundreds of years. By the early 1700s, several books included illustrations of ancient plants. The first comprehensive study of ancient plant life was written by Swiss doctor Johann Jacob Scheuchzer. Published in 1709, it was titled *Herbarium Diluvianum*. Over a century later, French botanist Adolphe-Théodore Brongniart was the first to explore the relationship between extinct and living plants. In 1822, he published a paper on the classification and distribution of fossil plants, which he followed-up with a two-volume *Histoire des Végétaux Fossiles* (*History of fossil plants*, 1828–37). Known as the "father of palaeobotany", he organized fossil plants in order, related extinct plants to the living ones they most closely resembled and applied principles for distinguishing them. His findings also helped to persuade a sceptical public about the immense history of Earth and how new, increasingly complex organisms had emerged over time.

The Godzilla of fungi

In 1843, at a time when palaeobotany was still in its infancy, a mysterious fossil was discovered that would continue to baffle scientists for over 150 years. Its appearance was similar to gigantic tree trunks, measuring up to 9 metres in length and as much as 1 metre wide. It dates back to between 360 and 470 million years ago and would have been by far the tallest living thing in its day. They were initially misidentified as conifers and named *Prototaxites* (which loosely means "first conifer"). Scientists finally figured out what the giant creature was after a fossil discovery in Saudi Arabia in 2007. Researchers from the University of Chicago examined the levels of carbon isotopes and concluded that it was a fungus. The same conclusion was put forward 20 years earlier by Francis Hueber, of Washington's National Museum of Natural History, but it was met with resounding disbelief and scepticism. The question remains: why would a fungus grow to such a gigantic size? Some suggest it was a reproductive strategy to spread its spores over wider distances.

Extinct giants

You may have heard a lot about the extinction of some of the largest creatures to have roamed the Earth, but several large, magnificent trees have also died out during the planet's history. Fossils of the following trees were first recorded by scientists during the nineteenth century.

SCALE TREES

Lepidodendron, also known as "scale tree", is an extinct prehistoric tree that was one of the most abundant trees of the Carboniferous Period (300–360 million years ago). Their stem (not trunk) was mostly made up of green, soft tissues instead of wood. Scientists have likened it to a giant herb that could grow to over 50 metres! Forests of this tree were dense, with up to 2,000 trees per hectare – this was possible because they grew as a straight pole until maturity, at which point the trees began branching out. During the nineteenth century, amateur fossil collectors displayed specimens at fairs, claiming they were from the skin of prehistoric giant lizards and snakes.

RAINBOW WOOD

Araucarioxylon arizonicum trees grew to 59 metres tall with trunks measuring 3 metres in diameter. They came in a variety of colours; some showing red and yellow tones, while others appearing purple. Fossils have been found in Northern Arizona, where they flourished around 200 million years ago.

INSPIRING MYTHOLOGY

Fossils of *Araucarioxylon arizonicum* were part of the mythology of local tribes in Arizona. The Navajo believed they were the bones of a giant killed by their ancestors, while the Paiute saw them as spent arrow shafts from the thunder god.

DIAMOND SCALES

Sigillaria trees existed for around 100 million years and went extinct around 383 million years ago. Instead of bark, the *Sigillaria* was protected by green diamond shaped scales, and its base was made of densely packed leaf bases.

The Terra Nova Expedition

"Fossil leaves, wood and pollen… provide a basis for palaeoclimatic interpretations that are informing debate about the relative stability of the Antarctic ice sheets."

Paul Kenrick

In 1910, a team of British explorers led by Captain Robert Falcon Scott set sail for Antarctica. In February 1912, they found fossils of plants, which the expedition's chief scientist, Edward Wilson, described as "distinctly beech like". The expedition ended in tragedy when the team perished on their return journey. Their geological samples and diaries were later found, but without fossils resembling beech leaves. Wilson's diary entry was dismissed as misidentification. To mark the expedition's centenary, scientists revisited the records in 2012 and confirmed Wilson was correct – he had found fossilized leaves of the Southern Beech (*Nothofagus*). Research in Antarctica had revealed fossils of Southern Beech in 3- to 4-million-year-old sediments.

Continental drift

In 1912, German geophysicist Alfred Wegener was one of the first scientists to argue that geological and biological similarities between the Earth's continents were caused by these continents moving over geological time – an idea known as "continental drift". Possibly the most important fossil evidence supporting this theory was a plant called *Glossopteris*. Known as a woody, seed-bearing tree that became extinct about 200 million years ago, it is named *glôssa*, the Greek word for tongue due to its tongue-shaped leaves. The wide distribution of *Glossopteris* across the southern continents of South America, Australia, Africa and Antarctica at the same point in the fossil record support Wegener's argument that these separate continents were once joined around 250 million years ago (in one "supercontinent" named *Gondwana*, which also included India). The *Glossopteris* seed was large and bulky, so it could not have drifted or flown across the oceans to separate continents.

Changing Earth's story

The Australian geologist Reg Sprigg was searching for minerals in Ediacaran Hills, Southern Australia, in 1946 when he encountered some fossil imprints that predated the Cambrian geological age (540–490 million years ago). The scientific community disagreed with this dating – they believed that simpler forms of life must have existed before the Cambrian Explosion but that no fossils survived.

In 2004, the Ediacaran Period, the first new geological era in more than 100 years, was recognized. The organisms discovered by Sprigg turned out to be 600–540 million years old (even older than the Burgess Shale) and turned accepted ideas about the Earth's history upside down: now there was proof of life *before* the Cambrian Explosion. They record the first known multicellular animal life on Earth that predated the Cambrian Period. The extraordinary organisms exposed by Sprigg tell us much about our geological heritage.

UNIQUE ORGANISMS

The discovery of these earliest fossils created a surge of interest in the Ediacaran and the Proterozoic Eras. However, how to characterize the fossils remains a challenge. The soft-bodied fossil impressions found in the Ediacara Hills were so hard to characterize that they became known as the Ediacaran biota rather than "flora" or "fauna". They resemble flatworms, soft corals and jellyfish and range in size from a few centimetres up to a metre long. They were often classified according to shape instead of any evolutionary relationships. Many of the organisms found at Ediacara may represent early algae, lichens or even multicellular "experiments", unlike anything else found in the world today. Palaeontologists continue to explore questions about these ancient organisms, including how they evolved, how they lived and which creatures are their descendants (if any). This discovery also gave scientists a better understanding of how fossils of organisms with soft tissue can become preserved in the fossil record.

The oldest forest

"Like discovering the botanical equivalent of dinosaur footprints."

William Stein

Fossils of the world's oldest trees, the Gilboa stumps, were first detected during excavations for the Gilboa Dam, New York State, in the 1850s and 1920. They were 385 million years old. In 1924, Winifred Goldring organized a museum exhibition that brought the "Gilboa forest" to global attention. Dating from the Devonian Period, the site became known as the "world's oldest fossil forest".

FEMALE PIONEER

In 1939, Winifred Goldring was appointed the first female State Paleontologist of the State of New York, making her the first woman in the U.S. to hold such a position. She also served as the first female president of the Paleontological Society in 1949.

A living fossil

In 1941, Japanese palaeobotanist Shigeru Miki discovered fossils of a tree that he had never seen described in the botanical literature. Believing that it was extinct, he published his research and named the tree *Metasequoia*. Shortly after, the Chinese forestry professor Zhan Wang was introduced to a spectacular tree in central China that was unknown to science. He discovered a shrine around the tree, and learned it was named *shui-sa*, or "water fir", by the locals due to its love of moist soil. Five years later, the Chinese palaeobotanist Hu Xiansu (or Hu Hsen-Hsu) looked at specimens of *shui-sa* and, recalling Miki's published description in 1941, recognized that it was a live *Metasequoia* – the tree was not extinct! This is considered one of the greatest botanical discoveries, not least because scientists on opposite sides of a war collaborated to start spreading the seeds worldwide. Xiansu worked with the director of the Harvard Arnold Arboretum (museum of trees) to plant samples of the tree in North America for the first time in 2 million years.

Ancient survivors

"The plants have continued to live on, while the dispersers themselves have already gone extinct."

Peter Crane

The *Ginkgo biloba*, otherwise known as the maidenhair tree, is unusually resistant to pollution, as well as cold temperatures, pests, fungi and diseases. This robustness also does not weaken with age, which is an incredible feat given they can live for thousands of years. They survived the asteroid that killed off the dinosaurs, and they remained standing in Hiroshima after the atomic bomb of 1945. What makes this hardiness surprising is that some of the traits of Ginkgo trees are usually considered weaknesses in volatile environments, such as their slow growth, large size and late reproductive maturity. In 1989 in China, palaeontologists unearthed 170-million-year-old Ginkgo fossils; the oldest found in the world to date. This discovery made them one of the oldest living tree species on Earth.

ANCIENT PLANTS MIRROR MODERN TREES

In 2003, Chinese palaeobotanists Zhou Zhiyan and Shaolin Zeng compared 121-million-year-old fossil material with modern Ginkgo. Aside from differences in the ways that the seeds are attached in the fossil plants, overall the species are very similar. Scientists have also compared old and young Ginkgos living today to find the secrets behind their longevity. The immune systems of 1,000-year-old trees are just like those of 20-year-old trees; they continue to make defences, growing wider and producing seeds indefinitely through old age.

Ginkgo is also a rare example of humans rescuing a species from the brink of natural extinction 11,000 years ago and spreading it around the world.

UNIQUE

Ginkgo is a single species with no living relatives – something unusual in the plant and animal world. The tree is also the only living connection between ferns and conifers.

Bountiful bacteria

The unsung heroes of Earth's natural history are tiny single-celled bacteria called *Cyanobacteria*. These photosynthetic organisms were the first life forms to produce oxygen 2.4 billion years ago, triggering the long process of creating the world. Fossils of some of the oldest specimens were found in 1965 in limestone in Bitter Springs, Australia. Over 850 million years old, they became known as blue-green algae, though this is inaccurate – *Cyanobacteria* do not have a nucleus; their genetic material mixes in with the rest of the cell, a characteristic associated with bacteria and archaea. The fossils also revealed that *Cyanobacteria* were surprisingly diverse from an early date. The Bitter Springs biota was one of the earliest successful demonstrations that fossils could be found in Precambrian rocks. These fossils, along with those found in the Gunflint chert in Canada (see page 58), constituted new evidence of early life.

Macroscopic discovery

In 2010, scientist Abderrazak El Albani and colleagues reported that 2.1-billion-year-old macroscopic fossils had been found in Gabon. The deposit showed multicellular life existed 1.5 billion years earlier than previously thought. This represented a key step in the evolution of life on Earth as the move from single-celled to multicellular organization allowed complex organisms to eventually emerge, including animals and plants.

EARLIEST LAND PLANTS

The appearance of plants on land was an important evolutionary breakthrough in Earth's history. In 2010, fossils of the oldest land plants ever found were revealed in Argentina. The liverwort specimens (simple plants lacking stems or roots) suggested that a diversity of land plants had evolved 472 million years ago. Discoveries made in subsequent years pushed the timeline back even further.

The first flowers

"A lot of paleontological progress happens by virtue of people seeing things in new ways."

Neil Shubin

There are over 350,000 species of *angiosperms*, or flowering plants, in the world today, but their origin used to be uncertain. In the nineteenth century, modern-looking *angiosperms* appeared suddenly in Late Cretaceous floras. Their abrupt arrival and rapid development puzzled palaeobotanists, with Charles Darwin famously calling it an "abominable mystery". The first Cretaceous flower unearthed by Bruce H. Tiffney in the U.S. was reported in 1977 but its discovery was considered a fluke. In 1981, the Danish palaeobotanist Else Marie Friis and her colleagues crumbled soft sediments into a sieve, washed away the sand grains in water and then used a microscope to examine the tiny specks of charcoal that remained (which were formed by natural fires).

The attempt proved successful: the remains of tiny three-dimensional flowers were found within the charcoal pieces, which turned out to be 80 million years old. They were no bigger than the full stop at the end of this sentence. This finding changed palaeobotanists' perception of ancestral flowers. Friis' breakthrough was finding the first of many charcoal flower fossils that showed *angiosperms* dating back to the early Cretaceous, or about 120 million years ago. Bigger flowers came later and were more advanced.

A LOST FRAGRANCE (AND TASTE)

Argocoffeopsis lemblinii is an extinct relative of the coffee plant. Its existence is only known from a description made by the French botanist Auguste Chevalier in 1907 while visiting the Ivory Coast. The flowers were white, and it grew in a forest habitat. It has not been seen since.

Oldest plant-like fossil

"It's science fiction, except that it's real."

Emma Hammarlund

In 2015, Swedish PhD student Therese Sallstedt spotted the fossilized remains of 1.6-billion-year-old red algae in a rock sample from from India. Before then, the earliest plants known were 600-million-year-old algae from China. Sallstedt's finding proved that complex life began much earlier than previously thought. The fossils were also unique in their preservation – even the microscopic internal structures of the organisms were visible.

THE TREE OF LIFE

The molecular clock, or "tree of life", indicates the time in the history of life on Earth when certain lineages diversified. This clock is anchored by dated fossils.

Response to climate change

In 2015, scientists first looked at the cells of ancient plants to measure the tree cover and density of trees, shrubs and bushes across different locations over time. The curves and sizes of cells indicate whether a plant grew in shade or sunlight. Researchers applied the method to 40-million-year-old *phytoliths* from Patagonia and discovered that habitats lost dense tree cover and spread out much earlier than previously thought. The decline in vegetation cover happened at the same time as cooling ocean temperatures and the evolution of animals with the type of teeth that feed in open, dusty habitats. Identifying the historic structure of vegetation provided a key insight into ecosystems. Palaeobotanists were now able to quantify in detail how Earth's plant and animal communities have responded to climate change over millions of years. As a result of this technique, scientists can now forecast how ecosystems will change under future climate scenarios.

Deep rooting

If you picture planet Earth, blue and green colours are likely to come to mind. However, until 450 million years ago, there was no life outside water, and the land surface was a rocky landscape rather than a lush green. One of the great transitions in Earth's history was this change in the landscape, and in 2016, scientists discovered plants with early root-like systems that originated 20 million years before the first forests grew. The fossils of the plant, called *Drepanophycus*, were excavated in Yunnan province, China by Jinzhuang Xue and his colleagues. They found roots of around 1 centimetre in diameter that branched continuously. They were tightly packed, with up to 1,000 roots in every square metre of sediment examined. The plant had a network of stems (rhizomes) which grew up, not down. The stems extended up through about 15 metres of sand and siltstone, leaving an underground network of dead rhizomes which would have helped stabilize the sediment. This ability to stabilize thick piles of sediment to make proto-soils was the first step in the evolution of deep-rooted trees.

A cradle of evolution

In 2018, the earliest records of three major plant groups were discovered near the Dead Sea in Jordan. The fossils were 255 million years old – 5 million years older than previously known. This was significant because it meant the plants had survived the world's greatest mass extinction, which occurred about 252 million years ago. The plant groups were:

- *Podocarpaceae* – the second-largest family of conifers today. This made the discovery the oldest fossil evidence from any living conifer family.
- *Corystospermaceae* – a group of seed plants that became extinct around 150 million years ago.
- *Bennettitales* – a peculiar lineage of extinct seed plants with flower-like reproductive organs.

The area would have had a tropical climate when the fossils formed.

Oldest green plant

Approximately 2 billion years ago, single-celled *Cyanobacteria* (see page 93) transitioned to multicellular plants. It paved the way for the riot of plants that inhabit the world today. The oldest fossils of a green plant known as *Proterocladus antiquus* were discovered in 2020. It was a green algae, no larger than a grain of rice, which carpeted areas of the seafloor about 1 billion years ago. It had many thin branches, which suggested that it attached itself to the seafloor with a root-like structure and thrived in shallow water. The discovery confirmed that green plants appeared at least 1 billion years ago, and they started in the ocean before they expanded their habitat to dry land.

The root of it all

The evolution of plants from simple stems to more complex forms with roots was transformative for our planet. Plant roots stabilized the soil, reduced atmospheric CO_2 levels and changed the circulation of water across the surfaces of continents. But, the question of *how* plants made this evolutionary step remained an enigma until a key insight was made in 2021. Scientists used digital techniques to produce a three-dimensional reconstruction of a 400-million-year-old fossilized plant found in Scotland, named *Asteroxylon mackiei* – an early ancestor of clubmosses. The model revealed that, unlike today's plants (that grow roots from within), the roots of *Asteroxylon* grew through "dichotomous branching". This is a process whereby the tip of a leafy shoot splits to make two new branches; one maintains its shoot identity and the other grows into a root. This fundamentally different way of developing roots was the unseen evolutionary stage between the earliest (rootless) plants and living clubmosses that have roots.

Pompeii of prehistoric plants

Noeggerathiales were plants that lived around 251–325 million years ago. Although palaeobotanists had learned about their diversity and distribution, their anatomy was unknown. The mystery was finally resolved in 2021, after complete fossils preserved in volcanic ash were found in Wuda, Inner Mongolia. The 298-million-year-old specimens turned out to be advanced members of the ancestors of seed plants: they had the spore propagation mode of ferns and the vascular tissue of seed plants. They are now recognized as advanced tree-ferns that evolved complex cone-like structures. This meant they were more closely related to seed plants. They belonged to a sister group of seed plants, the former *gymnosperm* (see page 78).

Going further back

The search for the world's first flowering plants has been notoriously difficult (see pages 95–96). After decades of effort, progress was made in 2022 with two major discoveries.

OLDEST FLOWERING BUD

Scientists in China found a 164-million-year-old plant named *Florigerminis jurassica*. The fossil was 4.2 centimetres long and 2 centimetres wide. The plant's stem was connected to a fruit, a leafy branch and a flower bud around 3 square millimetres in size – a trio of data that is especially rare. It was the earliest example of a flower bud ever found. The specimen was the most convincing evidence yet that *angiosperms* (flowering plants) developed in the Jurassic Period, between 145 and 201 million years ago. Palaeontologist Xin Wang suggested they would have been uncommon at that time compared with *gymnosperms* (nonflowering plants) and geographically isolated.

FLOWERS BEFORE DINOSAURS

The fossils of *Phylica* flowers encased in sap were found to be 260 million years old. It was previously believed that the *Phylica* evolved about 20 million years ago. The plant is part of the *Rhamnaceae* family of more than 1,000 species living worldwide today (a family that was previously believed to be less than 100 million years old). Dr Tianhua He said the specimen allows scientists "to examine the speed of evolution of *Rhamnaceae* and – with a bit of caution – extrapolate to all flowering plants".

PREHISTORIC BLOSSOM

The largest flower ever found encased in amber was uncovered in 2023, after it had spent 150 years sitting forgotten in a museum collection. *Stewartia kowalewskii* was over 2 centimetres wide and almost 40 million years old.

Plant behaviour

Palaeontologists can sometimes infer behavioural characteristics from fossils. In 2023, 250-million-year-old fossil plants showed the first evidence of sleep movements (known as "foliar nyctinasty"). This occurs when plants fold their leaves at night and open again in the morning. The discovery was based on symmetrical insect feeding patterns (unique to plants with nyctinastic behaviour) found on Permian plants known as *gigantopterids,* which died out 252 million years ago. This meant that sleeping behaviour evolved independently in different plant groups and at various times during Earth's history.

THE MOVE TO LAND

In 2022, palaeobotanists learned that plants changed their vascular systems 400 million years ago to extract water more efficiently from soil. This allowed plants to survive droughts and spread from watery habitats to drier land.

PALAEOANTHROPOLOGY

Palaeoanthropology is the study of ancient humans through fossil evidence, including bones and other preserved features, such as footprints. It explores the history of human evolution, as well as the development of tools, early human behaviour, how the first social groups were formed and the growth of different cultures. This chapter introduces you to some of the greatest discoveries made and the fascinating insights they have revealed about Hominins – the lineage of bipedal apes that diverged from other African apes (including humans and our humanoid relatives who walked upright). The different species of archaic humans will be outlined as well as the evolutionary links between them. You will see how palaeoanthropology informs our understanding of human biology and our diversity today, and why every discovery made sheds more light on what it means to be human.

The Gibraltar Skull

It was mid-nineteenth-century fossil finders who first
established that skeletal remains belonged to a different,
extinct form of human. In 1848, an unrecognized adult
skull was found by workmen at a quarry in Gibraltar.
Its significance was not appreciated until 1856 when
naturalist Johann Fuhlrott identified similar remains in
the Neander Valley, Germany, which were 40,000 years
old. In 1864, this newly discovered species of human
was named *Homo neanderthalensis* and began to inform
debates about human evolution. *Neanderthals* were
initially believed to be an inferior human that inhabited
Europe before modern people. As the first extinct
human relatives known to science, their unearthing and
the scientific debate they inspired marked the start of
palaeoanthropology as a science.

Historic Homo sapiens

In 1868, human skulls and other skeletal remains were located at the Cro-Magnon rock shelter in France. Around 30,000 years old, they were the first fossils to establish the ancient roots of our own species – *Homo sapiens.* They led a physically tough life – the remains of an adult female showed she had survived a skull fracture. One skull had signs of a fungal infection and others had fused vertebrae in their necks indicating traumatic injury. The survival of these individuals implied the availability of group support and care.

STONE AGE SENSATION

In 1879, eight-year-old Maria de Sautuola found cave drawings and engravings of animals in Altamira, Spain. They ranged from 14,000 to 35,000 years old. Because they challenged all assumptions about prehistoric people, the paintings were dismissed as a forgery for years.

Java man

Dutch surgeon Eugène Dubois found the first *Homo erectus* fossil on the Indonesian island of Java in 1891. He named the species *Pithecanthropus erectus*, or "erect ape-man". With evidence of a small brain and a fully upright posture, Dubois argued that he had found the evolutionary "missing link" between apes and modern humans, which sparked hot debate among anthropologists. Eventually, similarities between *Pithecanthropus erectus* (Java Man) and *Sinanthropus pekinensis* (Peking Man – see page 115) led Ernst Mayr to rename both *Homo erectus* in 1950, placing them directly in the human evolutionary tree.

HOMO ERECTUS

Homo erectus is by far the longest surviving of all our human ancestors, with fossil evidence stretching across 1.5 million years. They shared many similarities with modern humans, including body proportions and size, though their brains were much smaller. They were the first human species to make hand-axes.

Mauer jaw

In 1907, an enormous jawbone was discovered by a worker in Mauer, Germany. It was originally dismissed as being too ape-like for human ancestry, but the next year it was declared by anthropologist Otto Schoetensack as belonging to a new species. At the time, it was the oldest human jaw in the European fossil record at 640,000 years old. The fossil became the type specimen for *Homo heidelbergensis*, a species known as the most recent common ancestor between modern humans (*Homo sapiens*) and *Neanderthals*. In 1921, a 200,000–300,000-year-old skull unearthed in Zambia indicated that the species had a huge face, eyebrow ridges and a low forehead. During the following decades, palaeoanthropologists debated whether *Homo heidelbergensis* could be extended to all Middle Pleistocene (or Chibanian age) humans across the world, or if they were restricted to Europe. This confusion about the species-level classification of archaic human remains from 126,000 to 770,000 years ago is often known as the "muddle in the middle".

Missing links

For most of the twentieth century, it was believed that there was a single lineage leading to modern humans. It was ingrained in popular culture by Rudolf Zallinger's 1965 *March of Progress* illustration, which showed six figures in profile, starting on the left with a long-armed crouching ape and ending on the right with an upright human. Ape-men were considered the missing links within this chain.

FILLING IN THE GAPS

Many missing-link fossils have been revealed to be hoaxes. In 1911, collector Charles Dawson supposedly discovered "Piltdown Man", the remains of a large-brained human skull and an ape-like lower jaw. Scientists realized it was fake 41 years later: in fact, it was the lower jawbone of an orangutan combined with the skull of a modern human.

One of the best known "missing links" is *Australopithecus africanus*, first located in South Africa in 1925. Anatomist Raymond Dart, who first analyzed it the "Taung skull" fossil, claimed it was a direct ancestor of modern humans. Many more fossils of this species, dating from 2 to 2.5 million years old, have been found since. Various aspects of the skeleton showed it was bipedal, while possessing curved and ape-like fingers, which suggested the ability to climb trees. Males measured up to 140 centimetres and females up to 115 centimetres.

In the twenty-first century, the term "missing links" is no longer considered to be accurate. As paleoanthropologist Ian Tattersall has said, "We now know that the picture was much more complex than that, with a lot of now-extinct species jostling for ecological space and evolutionary success". This means *Australopithecus africanus* is not a human ancestor but one of many branches on the *Hominin* evolutionary tree.

Peking Man

In 1921, Swedish geologist Johan Gunnar Andersson entered a cave full of "dragon bones" near Beijing, China. These bones turned out to be the remains of an early human from about a million years ago. It was eventually named *Sinanthropus pekinensis* ("Chinese man of Peking"). In the 1950s, these specimens, and the ones from Java (see page 111), were placed in the new species of *Hominin, Homo erectus.* The Zhoukoudian cave became the most productive *Homo erectus* site in the world, with at least 12 individuals excavated, plus evidence of stone tools and the lighting of fires. From these specimens, it was clear there was *Hominin* activity in Northeast Asia hundreds of millennia earlier than previously thought. As a result, the Chinese archaeologist Lin Yan (mistakenly) claimed that Chinese people were "the Earth's most ancient original inhabitants". The discovery was instrumental to the development of anthropology in China.

The Cradle of Humankind

The Cradle of Humankind, a UNESCO World Heritage Site in South Africa, is the largest concentration of human ancestral remains in the world. The area includes Sterkfontein Caves, where the first adult *Australopithecus,* an ancient *Hominin,* was discovered in 1936. The fossil was estimated to be 2.3 million years old and helped corroborate the 1924 discovery of the Taung skull (see page 114), which turned out to be a juvenile *Australopithecus.* In 2022, new dating techniques revealed that some of the Sterkfontein artefacts may be a million years older than previously thought. Investigators ascertained that all the *Australopithecus*-bearing cave sediments date from about 3.4 to 3.7 million years old, rather than 2 to 2.5 million years old as formerly believed. That age placed the fossils towards the beginning of the *Australopithecus* era, rather than near the end. Accurate dating allows palaeoanthropologists to understand how and where humans evolved.

The "first family" of palaeontology

During the second half of the twentieth century, spectacular finds in Africa expanded the hominid family tree (see page 118). In particular, discoveries by members of the Leakey family of Britain transformed our understanding of early humans – the following are highlights:

1959

Remains of *Paranthropus boisei* found by Mary Leakey in Olduvai Gorge, Tanzania. A distant cousin of ours, rather than a direct ancestor, the fossil was 1.8 million years old.

1964

Homo habilis (meaning "handy man") announced as a new species after remains were discovered in Olduvai Gorge in 1960 by Mary and Louis Leakey. This primitive species, along with *Homo rudolfensis*, is one of the earliest members of the genus *Homo*.

The announcement of *Homo habilis* was a turning point in palaeoanthropology. It shifted the search for the first humans from Asia to Africa and began a debate about the origin of *Homo* that endures to this day.

1984

Richard Leakey uncovered a near-complete *Homo erectus* skeleton at Lake Turkana, Kenya. Nicknamed "Turkana Boy", the fossil was approximately 1.5 million years old and is the most complete fossil skeleton of a human ancestor ever found.

1994

Meave Leakey found fossils of a *Hominin* that walked upright 4.1 million years ago. It was named *Australopithecus anamensis.*

These discoveries strengthened the case that *Hominins* originated in Africa, and showed that the evolution of humans was more complex than scientists had previously believed.

Lucy

In 1974, palaeoanthropologists Dr Donald Johanson and Tom Gray, digging in Hadar, Ethiopia, found the partial skeleton of the earliest known hominid thus far. They called her Lucy because the Beatles' song "Lucy in the Sky with Diamonds" was playing as they celebrated the discovery. At nearly 3.2 million years old, she was a member of the species *Australopithecus afarensis*, which lived from 2.9 million to 3.9 million years ago.

Lucy's brain and body was the size of a chimpanzee's. She was a fully mature adult female, standing at just a metre tall. But her ankle, knee and pelvis showed that she walked upright. This indicated that hominids walked upright long before they developed large brains and before the use of stone tools. Lucy also had powerful arms and long, curved toes that suggested an ability to climb trees.

The ancestral split

In 1948, French palaeontologist Camille Arambourg introduced the term "*Hominini*" to refer to humans and all our extinct bipedal ancestors. Hominids are all species of modern and extinct Great Apes originating after the human/African ape ancestral split, including *Homo* (of which only *Homo sapiens* – modern humans – remain), chimpanzees, gorillas, orangutans and all their immediate ancestors.

At the time of being unearthed, Lucy was the oldest and most complete early *Hominin* skeleton ever found. She supported the scientific view that human evolution was a gradual process involving the appearance and survival of transitional forms over millions of years. As Dr Johanson said, the *Australopithecus afarensis* species was not the "missing link" between apes and humans, but "one of the important evolutionary intermediaries between more ancient, more ape-like creatures and more recent, more [modern] human-like ancestors". Lucy's Ethiopian name is Dinkinesh, which translates to "you are marvellous".

A branching bush

Until the 1970s, many scientists believed that only one *Hominin* species could occupy an area at any one time and that there was one single evolutionary line leading from them to modern humans. This theory was upended by the discovery of Lucy and another fossil a year later. In 1975, Bernard Ngeneo and Richard Leakey discovered something in Koobi Fora, Kenya, which was initially considered to be *Homo erectus*, but many now classify it as *Homo ergaster.* The fossil was found in the same layer as a specimen of *Paranthropus boisei*, proving that human evolution was more like a branching bush than a linear tree.

SCEPTICAL SCIENTISTS

Between 1953 and 1968, Dutch archaeologist Theodor Verhoeven found stone tools on the Indonesian island of Flores. He claimed they were made by *Homo erectus* around 750,000 years ago – long before modern humans arrived there. Only in the 1990s did further research validate his theory.

Ardi

What came before Lucy? When, and how, did an upright posture develop? In 1994, these questions were answered. A skeleton of a female who lived 4.4 million years ago was located in the Afar desert, Ethiopia. Belonging to the species *Ardipithecus ramidus* and nicknamed "Ardi", the fossil transformed basic ideas about how our earliest ancestors looked and moved. Though she is not the oldest member of the extended human family discovered to date, she is the most complete early hominid found: 125 pieces of Ardi's skeleton were located. After 15 years of in-depth analysis, scientists reported that Ardi was unlike a chimpanzee, gorilla or human. She had a combination of traits that allowed her to live both on the ground and in the trees. This discovery proved that humans did *not* evolve from knuckle-walking apes, as previously believed.

The earliest hominids

"The debate is open, even between members of our team."

Franck Guy

Breakthroughs in the early twenty-first century changed ideas about the split between the human and chimp evolutionary line. In 2001, hominid fossils dated between 5.8 and 6.2 million years old were found in Kenya. The new species was named *Orrorin tugenensis.* Further analysis confirmed it could walk upright. In 2002, *Sahelanthropus tchadensis,* a 7-million-year-old fossil found in Chad, was a possible ancestor of both chimps and humans. In 2022, studies suggested that it both walked upright on the ground and climbed trees. Its relationship to humans is still undecided.

Interbreeding

The Swedish geneticist Svante Pääbo sequenced the genome of *Neanderthals* in 2009. The following year, he discovered a new *Hominin* species, *Denisovans*, from DNA extracted from a finger bone found in Siberia. The fragment was 30,000 to 50,000 years old and evinced that a third human species was still in existence during this time (alongside *Neanderthals* and *Homo Sapiens*). Pääbo also promulgated evidence of interbreeding between human species. His work demonstrated that *Homo sapiens* had interbred with both *Denisovans* and *Neanderthals* after migrating out of Africa.

In modern humans, people of African descent have close to 0.5 per cent Neanderthal DNA in their genome, while people of European or Asian descent have roughly 1.7 and 1.8 per cent, respectively.

The percentage of *Denisovan* DNA in modern humans is highest in the Melanesian population (4–6 per cent), lower in other Southeast Asian and Pacific Islander populations, and very low or undetectable elsewhere. Some of the genes from our extinct relatives affect our immune systems during infections. A gene variant believed to have originated from the *Denisovans* and now common in Tibetan people protects against the physiological effects of high altitude. Moreover, the work of Pääbo and Hugo Zeberg has also revealed that genes inherited from *Neanderthals* can affect our fertility and pain thresholds.

THE HYBRID DAUGHTER

In 2018, DNA studies by Svante Pääbo's team showed that a 90,000-year-old bone fragment found in the Altai Mountains, Russia, came from a female who had a Neanderthal mother and *Denisovan* father. She was nicknamed "Denny",

A common ancestor

Our closest ancient human relatives are *Denisovans* and *Neanderthals*. Potentially, modern human lineages and *Neanderthals* split from a common ancestor roughly half a million years ago. The identity of this common ancestor continues to be debated. For decades, it was considered to be a species called *Homo heidelbergensis* – a fossil found in Zambia in 1921 was believed to be 500,000 years old. But, technological developments in 2020 revealed it was 299,000 years old; too recent to be the common ancestor. A more plausible contender is *Homo antecessor*, known from specimens found in Gran Dolina, Spain. Between 1994 and 1996, researchers found fossils belonging to six hominid individuals that lived roughly 800,000 years ago. They were the oldest human remains found in Western Europe to date. Their teeth were primitive, like those of *Homo erectus*, with facial features that resembled those of modern humans. However, most of the specimens represented children, whose features could have changed upon reaching adulthood. Complete adult fossils need to be found before scientists can confirm their physical appearance.

Rewriting family history

In 2020, the earliest known skull of *Homo erectus* was discovered in South Africa. It was 2 million years old and broken into 150 individual fragments. *Homo erectus* marks an important stage in human evolution when our ancestors began to accumulate adaptations that made them increasingly human-like. This fossil showed that *Homo erectus* existed 100,000–200,000 years earlier than previously determined and that it evolved in Africa. The specimen's brain was smaller than those of later *Homo erectus* fossils but larger than the more ape-like early humans *Australopithecus*. *Homo erectus* shared the landscape with two other types of humans in South Africa, *Paranthropus* and *Australopithecus* – meaning that *Australopithecus sediba* may not have been the direct ancestor of *Homo erectus*, or *Homo sapiens*, as previously believed.

Dragon man

In 2018, a massive fossilized skull was donated to the Hebei GEO University's palaeoanthropology department. It had been first found by a labourer in Harbin, China, in 1933. Japan was occupying the area at the time and the labourer, recognizing the fossil's significance (four years after Peking Man was discovered), buried it in an abandoned well to protect it from enemy forces. The labourer only told his family 85 years later, shortly before he died. When the family gave the skull to scientists, they were struck by how well preserved it was. The fossil was 146,000 years old, with its own distinctive combination of features that were unlike those of any hominid previously discovered.

The individual was an adult of great size. He had a wide face, large nose and mouth, and low, flat cheekbones that resemble modern people more than other extinct *Hominins*. The brain was 7 per cent larger than the average modern human brain. The new species, *Homo longi*, was nicknamed "Dragon Man" for the region where he was found. The discovery could potentially change views of how, and where, *Homo sapiens*, evolved.

MOVE ASIDE EUROPE AND AFRICA!

Spectacular fossils have been discovered across Asia in the twenty-first century. In 2018, 2.1-million-year-old stone tools were found in China, suggesting *Hominins* lived outside Africa much earlier than previously thought. A new species named *Homo luzonensis* was found on Luzon island, the Philippines, in 2019. Dated between 50,000 years and 67,000 years old, the find raised questions about how *Hominins* travelled to the island.

Food and drink

Major breakthroughs were made in 2022 regarding the origins of modern food staples, including the following:

MEAT DID NOT MAKE US

The disproportionately large brains of humans first appeared in *Homo erectus* nearly 2 million years ago. Did meat consumption increase after the rise in the use of stone tools about 2 million years ago, triggering the expansion in our brain size, or is it just that there is more meat-eating evidence available from that period? The archaeological evidence for meat eating increases dramatically after the appearance of *Homo erectus*, but scientists argue that this is simply because this time period has been researched more. There is no strong relationship between eating more meat and the evolution of larger brains in our ancestors.

Scientists have concluded that intensive sampling, not a dietary shift (from a plant-based regime to meat), was likely to be the cause of all the evidence of meat eating. Therefore, meat eating may not have been a big trigger in our evolution.

FROM RAW TO COOKED FOOD

The earliest evidence of prehistoric humans using fire to cook food dates back to 780,000 years ago. Until this discovery, the oldest signs of cooking were approximately 170,000 years old. The transition from raw food to cooked food reduced the energy required to break down and digest food, allowing other physical features to develop, including changes in the structure of the human jaw and skull.

CHICKEN DOMESTICATION

The origin of the domestic chicken has been traced to around 3,500 years ago in Thailand. Chickens then appear to follow grains (specifically rice and millet) as they spread around the world as a food source.

First surgeons

It has long been thought that the medical expertise of foraging communities like hunter-gatherers was basic and unchanging. But in 2022, palaeoanthropologists found the skeleton of a young adult whose lower left leg had been amputated in childhood by a prehistoric surgeon 31,000 years ago – long before the dawn of agriculture. It was the earliest evidence of a successful amputation and complex medical knowledge. Researchers concluded that the surgeon who treated this patient:

- Made use of a "natural pharmacy": the tropical rainforests of Borneo likely provided medicinal plants before, during and after the procedure (including anaesthetics and antimicrobial remedies preventing infection).

- Relied on community care: support was needed for the patient to survive a childhood amputation and to live for another six to nine years.

A Neanderthal community

Researchers identified a Neanderthal community for the first time in 2022, thanks to DNA extracted from bones. The fossils were of a father and his teenage daughter, plus others who were close relatives. They lived in Siberian caves around 54,000 years ago and used stone to create tools. Scientists found little genetic diversity within the clan, suggesting that *Neanderthals* here lived in small groups of 10 or 20 individuals – like the group sizes of endangered species on the verge of extinction. Women may have moved between communities more than men, possibly for mate selection.

SCULPTURE GARDEN

In 2016, religious or cultural artefacts by *Neanderthals* were found in the hidden depths of a cave in France. The bizarre circle sculptures made of stalagmites (a type of rock formation) were over 175,000 years old. This may be the oldest work of art on Earth.

Social groups

The 125,000-year-old bones of about 70 elephants were found in the 1980s in a German coal quarry. In 2023, scientists found that the elephants, which were three times the size of Asian elephants today, had been hunted by *Neanderthals*. The huge quantity of food provided by just one 10-tonne elephant would have sustained a group of *Neanderthals* for months. It also proved that they lived in large social groups, otherwise the precious meat would have become rotten before it could be consumed.

ROOTS OF COMPASSION

Evidence of group care and compassion has been found as far back as 1.77 million years ago in Dmanisi, Georgia. An elderly *Homo erectus* man survived with one tooth for years before his death. Members of his social group must have taken care of him.

Making modern humans

(MYA = million years ago, YA = years ago)

TIME	EVENT
55 MYA	First primitive primates evolve.
8–6 MYA	First gorillas evolve. Later, chimp and human lineages diverge.
6 MYA	*Orrorin tugenensis*, oldest human ancestor thought to have walked on two legs (bipedalism).
5.6 MYA	*Ardipithecus*, the first forest-dwelling bipedal apes appear.
5–2 MYA	*Australopithecines*, referred to as "man-apes" exist. Brains no larger than a chimpanzee's, but increasingly bipedal.
3.3–3.2 MYA	First Oldowan (stone tool industry) tools created by striking stones together. Lucy, famous specimen of *Australopithecus afarensis*, lives near what is now Hadar, Ethiopia.
2.5 MYA	The Palaeolithic/Old Stone Age Period begins. Brain expansion at this stage.

2 MYA	Human-like *Hominins* appear (*Homo erectus*, also referred to as *Homo ergaster*).
2–1.5 MYA	*Homo erectus* migrates out of Africa in large numbers (to Asia and later Europe); first hunter-gatherer ancestor and use of fire.
600,000– 200,000 YA	*Homo Heidelbergensis* lives in Africa and Europe (and possibly Asia); similar brain capacity to modern humans. *Neanderthals* appear 400,000 YA.
200,000 YA	*Homo sapiens* (anatomically like modern humans) appear in East Africa. Genetically, we can all trace our lineage back to this small group of humans – separated by approximately 6,666 generations. *Denisovans* exist at this stage.
150,000 YA	Possibly capable of speech; 100,000-year-old shell jewellery suggests the development of complex speech and symbolism.
117,000 YA	*Homo erectus* die out.
100,000 YA	*Homo sapiens* start migrating out of Africa; exchange of genes with other human species already present in Europe and Asia.

65,000– 50,000 YA	"Great leap forward" in several locations around the world: designing decorative jewellery and musical instruments; new weapons and hunting technology invented; cave walls adorned with art and figurative sculptures created – all requiring skill, foresight and abstract thinking.
40,000– 35,000 YA	*Neanderthals* die out.
30,000 YA	*Denisovans* die out. *Homo sapiens* become the sole surviving member of the human family.
12,000 YA	Agriculture develops and spreads; first villages.
6,000– 5,000 YA	The Sumerians of Mesopotamia (present-day Iraq) develop the world's first civilization.
5,500 YA	Stone Age ends and Bronze Age begins. Humans start to smelt and work copper and tin, and use them instead of stone implements.
5,500 YA	Earliest known writing.

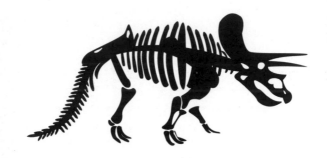

CONCLUSION

In the early nineteenth century, when palaeontology was still in its infancy, Lord Byron wrote: "truth is always strange; Stranger than fiction: if it could be told... How differently the world would men behold!" As this book has revealed, though you belong to the species that currently dominates Earth, the story that led to our present moment in the limelight is filled with multiple, parallel evolutionary paths spanning billions of years and an extraordinary diversity of organisms. The world as you see it today, and humanity's position in it, was far from inevitable. Studying ancient fossils allows us to imagine an alternative world encompassing alien-like organisms, magnificent beasts and spectacular landscapes – a real world that connects you physically through your existence but separates you by vast tracts of time. Palaeontology is not just about peering into the past: it reminds us that our existence is equally precarious but, also, exceptional. Unlike our ancestors, we are aware of the possibilities of extinction, evolutionary change and biodiversity – and we can choose – or not – to protect the planet's inhabitants for millennia to come.

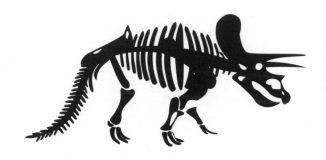

FURTHER
READING

Michael J. Benton, *The Dinosaurs Rediscovered: How a Scientific Revolution is Rewriting History* (2020)

Steve Brusatte, *The Rise and Reign of the Mammals: A New History, from the Shadow of the Dinosaurs to Us* (2022)

Henry Gee, *A (Very) Short History of Life on Earth: 4.6 Billion Years in 12 Chapters* (2022)

Dr Thomas Halliday, *Otherlands: A World in the Making* (2022)

Tom Higham, *The World Before Us: How Science is Revealing a New Story of Our Human Origins* (2021)

Dr Elsa Panciroli, *The Earth: A Biography of Life: The Story of Life on Our Planet Through 47 Incredible Organisms* (2022)

Rebecca Wragg Sykes, *Kindred: Neanderthal Life, Love, Death and Art* (2020)

THE LITTLE BOOK OF
ANTHROPOLOGY

Rasha Barrage

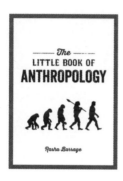

Paperback
ISBN: 978-1-80007-415-6

This illuminating little book will introduce you to the key thinkers, themes and theories you need to know to understand the development of human beings, and how our history has informed the way we live today. A perfect gift for anyone taking their first steps into the world of anthropology, as well as for those who want to brush up their knowledge.

THE LITTLE BOOK OF
PHILOSOPHY

Rachel Poulton

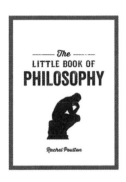

Paperback
ISBN: 978-1-78685-808-5

If you want to know your Socrates from your Sartre and your Confucius from your Kant, this approachable little book will introduce you to the key thinkers, themes and theories you need to know to understand how human ideas have sculpted the world we live in and the way we think today.

Have you enjoyed this book? If so, find us on
Facebook at **Summersdale Publishers**, on Twitter at
@Summersdale and on Instagram and TikTok
at **@summersdalebooks** and get in touch.
We'd love to hear from you!

www.summersdale.com

Image Credits
Cover and throughout – triceratops
© Magura/Shutterstock.com
Cover and throughout – ammonite and leaf
© intueri/Shutterstock.com